Mineral Identification Made Easy

Books by Patrick Nurre. These are all also available with sample rock, mineral, and fossil kits from Northwest Treasures at NorthwestRockAndFossil.com.

Rocks and Minerals for Little Eyes (PreK-3)
Fossils and Dinosaurs for Little Eyes (PreK-3)
Volcanoes for Little Eyes (PreK-3)
Geology for Kids (3-8) (Activity journal also available)
Rock Identification Made Easy (3-12)
Fossil Identification Made Easy (3-12)
Mineral Identification Made Easy (5-12)
Bedrock Geology (high school)
Rocks and Minerals: The Stuff of the Earth (high school)
Volcanoes, Volcanic Rocks and Earthquakes (high school)
Fossils Dinosaurs and Cave Men (high school)
The Geology of Yellowstone – A Biblical Guide
Genesis Rock Solid – A Biblical View of Geology
Geology and the Hawaiian Islands
Geology and Our National Parks

Mineral Identification Made Easy

Patrick Nurre

The Northwest Treasures Curriculum Project

Mineral Identification Made Easy
Published by Northwest Treasures
Bothell, Washington
425-488-6848
NorthwestRockAndFossil.com
northwestexpedition@msn.com

Cover photo: Vicki Nurre
Title page photo: By Didier Descouens - Own work, CC BY-SA 3.0,
https://commons.wikimedia.org/w/index.php?curid=6911957

Contents

How to use this book

Whether you are using this book for pleasure or a class credit, it is recommended that you take the necessary time to *master* the principles in each lesson before going to the next lesson. This is crucial in order to understand the nature of minerals.

The grade levels recommended for this study are 5-12.

Most lessons start with a word bank. Not all of the definitions for the words in the word bank are found in the text. You will need to look up some of these. Look these terms up before you start your study, and record what you discover in your notebook.

Most of the activities for each lesson center around the minerals and rocks that you will be studying, so it is important that you have a set of samples that you can refer to. The samples you will need are on the materials list. If you do not have these samples but would like to order a set, contact me at northwestexpedition@msn.com.

Materials List

The following samples and items are needed to complete this study:

- Storage box for your rocks and minerals
- Notebook for notes, observations, and vocabulary
- 12 rock-forming minerals including quartz, muscovite mica, potassium feldspar, sodium feldspar, calcium feldspar, biotite mica, amphibole (hornblende var.), pyroxene (augite var.), jasper, magnetite, calcite, and olivine
- Nine Mohs minerals including talc, gypsum, calcite, fluorite, apatite, feldspar, quartz crystal, topaz and corundum
- 10 minerals illustrating crystal habits including a beryl crystal, fluorite crystal, ruby crystal, calcite crystal, barite rosette, galena crystal, gypsum rosette, magnetite crystal, garnet crystal
- Set of representative minerals for each of the mineral families: silicate family, native element, oxide family, carbonate family, halide family, sulfate family, sulfide family, phosphate family, borate family (9 minerals total)
- The Mohs hardness test kit including instructions, hardness table, white and black streak plates, glass scratch plate, card-stock test strips, magnet and magnifying glass
- 11 basic rock types illustrating their most common characteristic. The set should include 2 plutonic rocks (granite and gabbro), 2 volcanic rocks (basalt and rhyolite), 2 pyroclastic rocks (pumice and tuff), 2 metamorphic rocks (gneiss and quartzite), and 3 sedimentary rocks (sandstone, limestone, and bituminous coal)
- 10 gemstones (in the rough or polished): suggested gemstones – agate, sapphire, emerald, tourmaline, iolite, lapis lazuli, green garnet, aquamarine, lace agate, amethyst

If you purchased the book separately from the kit and you would like to order the kit, please contact us at northwestexpedition@msn.com.

Have a fun experience as you collect the beautiful minerals all around us. If you ever have questions about this wonderful hobby, please don't hesitate to contact me.

Patrick Nurre
February 6, 2019

Lesson One – What is a Mineral?

Word bank: *cryptocrystalline, mineral, precipitate*

Minerals are often confused with rocks. And sometimes they are not easily seen to be different. So, let's begin with minerals. We will talk about the difference between rocks and minerals in a later lesson.

What is a Mineral?

The typical dictionary definition reads something like this: "A solid homogeneous crystalline chemical element or compound that results from the inorganic processes of nature."[1]

But there are a few challenges with this definition:

1. It is not easy to remember.
2. Most minerals have both a crystalline and a cryptocrystalline structure. The word cryptocrystalline meaning *hidden crystal*, refers to the fact that the normal crystal structure typically identified with that mineral is not visible to the naked eye. Because most minerals have both of these structures, it is often difficult to identify a mineral.
3. There are a few minerals which are also sedimentary rocks, such as halite and gypsum, and they have the same chemical formulas as their mineral counterpart.
4. We don't know the origin of minerals.

Let's see if we can simplify this definition. ***"Minerals are arrangements of elements. It is these elements that when combined in special ways, give us a unique substance with a particular shape and chemical identity. When these minerals are combined together, they form rocks."*** I know it doesn't sound very scientific, does it? But it does help us associate minerals with the rocks all around us. We will enlarge this definition as we progress in our study.

By the way, most of the minerals you will be concerned with are those that make up the rocks. And most of the isolated minerals you will find will be of the cryptocrystalline variety unless you find a special fee-based mineral collecting site.

---

[1] https://www.merriam-webster.com/dictionary/mineral

Now to the previous simple definition we can add the following:

1. **Minerals have a specific chemical formula.** For example, the mineral quartz is made up of the chemical formula, **SiO_2**, the elements silicon and oxygen arranged in a very special way to give us quartz. *Don't worry if chemical formulas are foreign to you. We will talk more about it in a later lesson.*

A beautiful lemon quartz crystal from Arkansas

Other minerals have much more complicated chemical formulas such as hornblende which is a type of amphibole. Its chemical formula is $Ca_2(Mg,Fe)_4Al(Si_7Al)O_{22}(OH,F)_2$, the elements calcium, sodium, magnesium, iron, aluminum, silicon, oxygen, hydrogen and fluorine all arranged in a very special way to give us the dark-colored mineral, hornblende.

The mineral hornblende – a black or dark green mineral shaped in needle-like structures

2. Minerals have both a **specific crystal structure** and a *cryptocrystalline* structure.

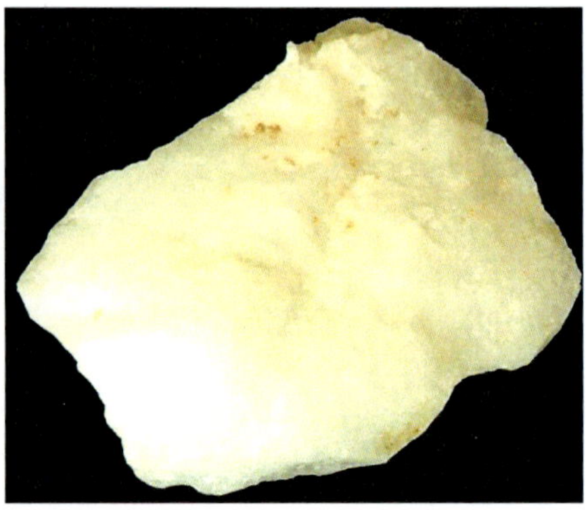

A quartz crystal always has six sides and cryptocrystalline quartz has an amorphous shape.

3. Some minerals are **both** rocks and minerals. Here are two common examples:
    - **Halite** – sodium chloride (salt); made of the elements sodium and chlorine. It is both a mineral and a sedimentary rock. It is a precipitate, meaning that it is a deposit from super saturated water.

Common halite occurs as groups of blocky shapes.

    - **Gypsum** – Calcium sulfate; made of the elements calcium, sulfur, oxygen and hydrogen. Gypsum occurs in many different shapes, but always has the same basic chemical formula. It is both a rock and a mineral. Like halite, it is also a precipitate.

Gypsum occurring as blades

(Left) Gypsum occurring as rosettes
(Right) Gypsum occurring as sheets of clear fibrous selenite called *satin spar*

Gypsum occurring as white cryptocrystalline veins in
Caprock Canyon State Park, Texas

Gypsum occurring as alabaster, a soft form of orange gypsum ideal for carving

4. In addition to exhibiting several different structural patterns, the same mineral can exhibit several different colors due to the addition of other elements that are called impurities. For example, the addition of iron usually makes things red or brown. Other colors are highlighted by applying heat or irradiation.

(Left) Smoky quartz crystal; some believe that the dark color is due to free silicon which is silicon dioxide colored by natural irradiation. But that is a guess. (Right) Rhombohedral calcite (a prism with six faces, each being a rhombus) colored by iron

Activity: Learn the difference between a rock and a mineral. In the following exercise, place an (m) underneath the mineral, and an (r) underneath the rock. Clue: Minerals can have specific shapes, and are often uniform in color. Answers are at the end of the pictures.

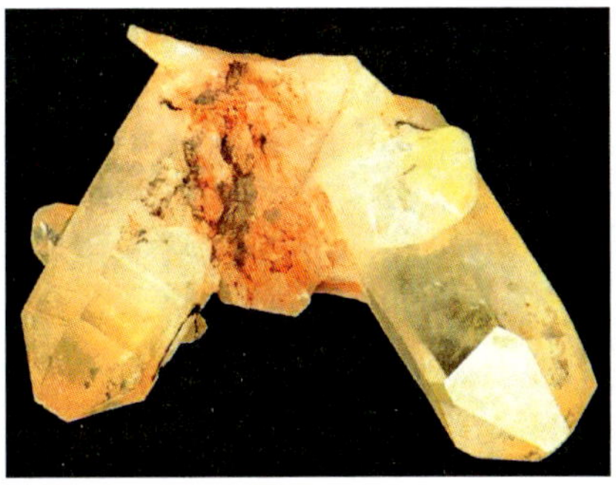

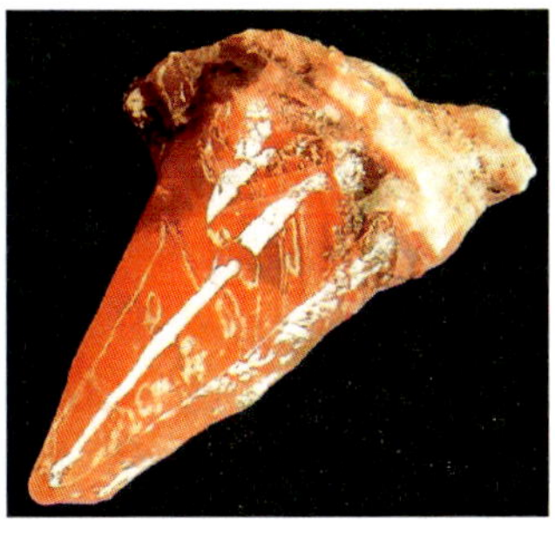

Answers: Row 1: m, r, r
Answers: Row 2: m, m, m
Answers: Row 3: r, r, m
Answers: Row 4: m, r
Answers: Row 5: r, r

Lesson Two – What are Minerals Made of?

Word Bank: atom, elements, periodic table

Now, to the rocks. Many confuse rocks with minerals, and vice versa. Quite simply, rocks are made of minerals. But what are minerals made of? When God created the space and the earth, in Genesis 1:1, He used a basic unit for all matter and energy – the ATOM. The word atom is from two Greek roots, *a*, without, and *tom*, cut; in other words, *without cutting*. The Greeks thought that matter could only be broken down (cut down) so far. They did not understand the atom as we know it today, but they were on the right track. The atom is made up of three particles called protons, neutrons and electrons.

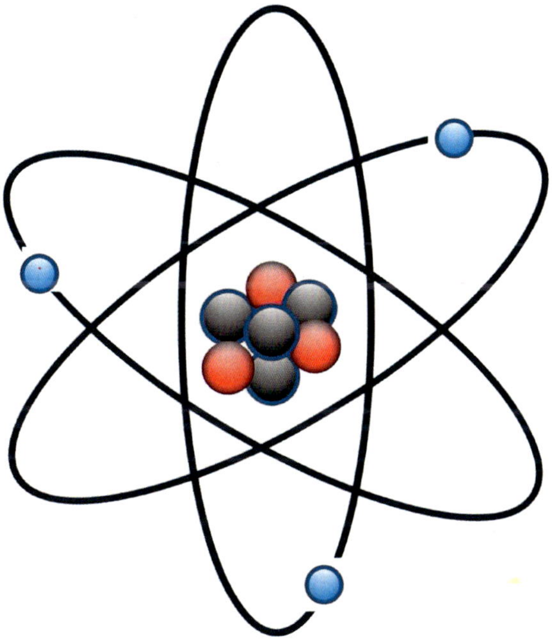

The atom – the central part of the atom, called the nucleus, is made up of protons and neutrons. The particles that swirl around the nucleus are called electrons.

God then used atoms to make ELEMENTS. Elements are arrangements of the various particles of atoms, the electrons, protons, and neutrons. A simple addition of one electron, proton or neutron changes the entire structure of the element and makes it into something totally different from its neighbor.

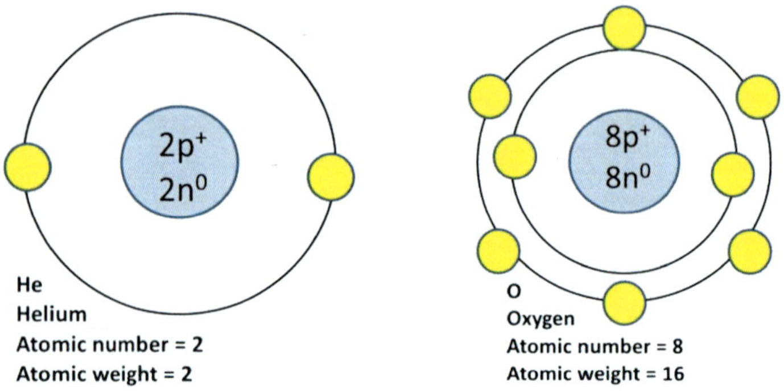

The element helium is made up of two protons, two neutrons and two electrons.
The element oxygen has eight protons, eight neutrons and eight electrons

The elements we know about to date are all arranged into a table based on the numbers of protons, electrons, and neutrons. Every element does something different. This table is called The Periodic Table and it is a great man-made arrangement of the various substances God used to make MINERALS.

Group→ ↓Period	1	2	3	4	5	6	7	8	9	10	11	12	13	14	15	16	17	18
1	1 H																	2 He
2	3 Li	4 Be											5 B	6 C	7 N	8 O	9 F	10 Ne
3	11 Na	12 Mg											13 Al	14 Si	15 P	16 S	17 Cl	18 Ar
4	19 K	20 Ca	21 Sc	22 Ti	23 V	24 Cr	25 Mn	26 Fe	27 Co	28 Ni	29 Cu	30 Zn	31 Ga	32 Ge	33 As	34 Se	35 Br	36 Kr
5	37 Rb	38 Sr	39 Y	40 Zr	41 Nb	42 Mo	43 Tc	44 Ru	45 Rh	46 Pd	47 Ag	48 Cd	49 In	50 Sn	51 Sb	52 Te	53 I	54 Xe
6	55 Cs	56 Ba	*	72 Hf	73 Ta	74 W	75 Re	76 Os	77 Ir	78 Pt	79 Au	80 Hg	81 Tl	82 Pb	83 Bi	84 Po	85 At	86 Rn
7	87 Fr	88 Ra	**	104 Rf	105 Db	106 Sg	107 Bh	108 Hs	109 Mt	110 Ds	111 Rg	112 Cn	113 Uut	114 Fl	115 Uup	116 Lv	117 Uus	118 Uuo
		*	57 La	58 Ce	59 Pr	60 Nd	61 Pm	62 Sm	63 Eu	64 Gd	65 Tb	66 Dy	67 Ho	68 Er	69 Tm	70 Yb	71 Lu	
		**	89 Ac	90 Th	91 Pa	92 U	93 Np	94 Pu	95 Am	96 Cm	97 Bk	98 Cf	99 Es	100 Fm	101 Md	102 No	103 Lr	

The Periodic Table – Dmitri Mendeleev, a Russian scientist is generally credited with the publication, in 1869, of the first widely recognized periodic table

God used various **minerals** as the basic building blocks of **rocks**, which God then used to make the earth.

The most abundant mineral on earth is quartz. Quartz is made up of two elements called silicon (Si) and oxygen (O) arranged in a very special way to give us a very special mineral. And what's so special about quartz? Quartz, to some degree, is in just about every rock of the earth!

Time for some memorization! I use the following illustration to help keep the various building blocks of rocks in order. Here is how it goes: *God used atoms as the basic building blocks of all matter and energy and He arranged these atoms into elements. He used elements to make minerals. And He used minerals to make rocks.*

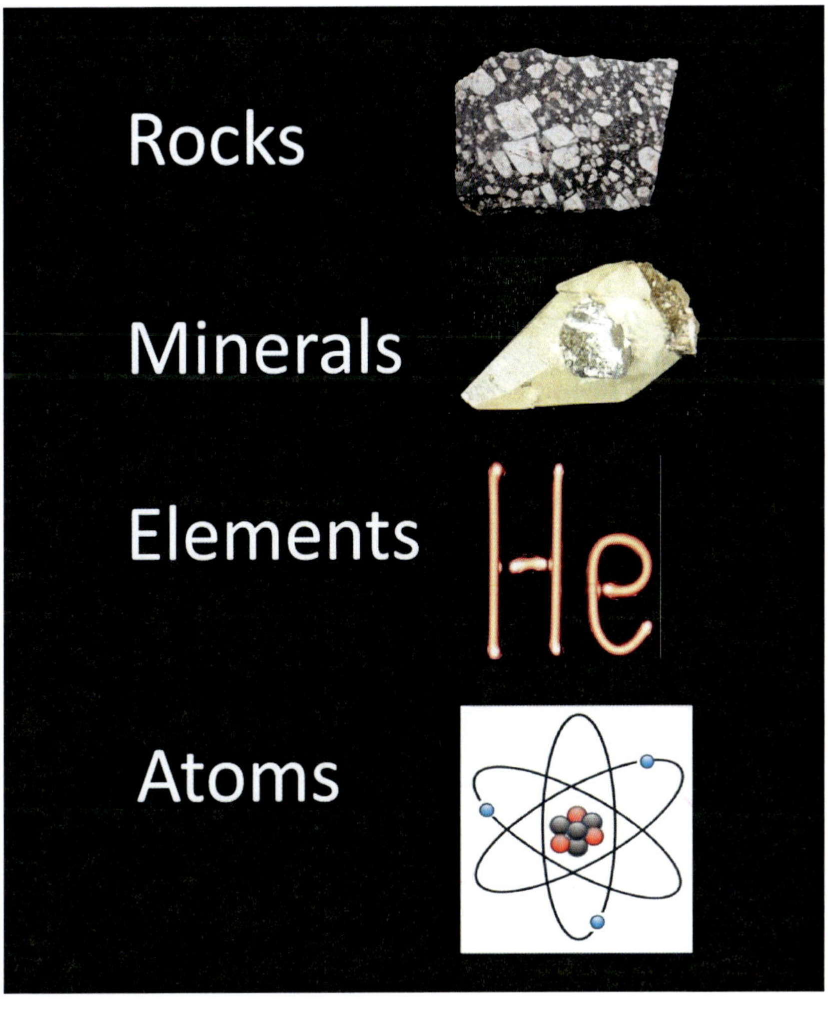

Activity: Memorize *The Building Blocks of Rocks* diagram and have someone quiz you on it.

Lesson Three – The Rock-Forming Minerals

***Word bank:** felsic, mafic,*

There are over 4,500 recognized minerals today! Remember that minerals are made up of elements and most of these elements are used in our bodies to help them to grow. Even the element arsenic - just a little bit, of course! - is needed in our body to be healthy. Blood that is low in iron is weak. Bones low in calcium will become brittle. Elements are in our cars, bathrooms, soap, and even in our medicines. Many of these same elements are also found in the minerals that make up the rocks all over the earth.

The earth is filled with elements. But an interesting and fascinating point is that most of the rock-forming minerals are made up of the eight most abundant elements of the earth's crust. You will learn what those rock-forming minerals are and their chemical structures in a just little bit. First, though, study the chart below and see if you can memorize these eight most abundant elements of the earth's crust.

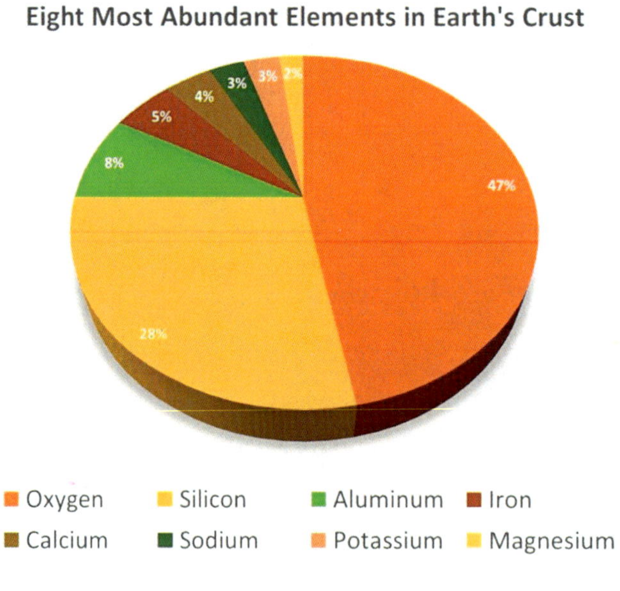

The other fascinating thing is that the two most abundant elements in the earth's crust are **oxygen** and **silicon**. What does that tell you? Well, quartz is made up of oxygen and silicon and it is found in just about every rock on earth. And it so happens that the largest mineral family, which we will learn about in Lesson Four, is the Silicate family – those minerals that contain the combination of **SiO** (**silicon** and **oxygen**) together! It is just amazing to see how God made the earth and everything in it to work in an orderly and useful manner.

Now, at this point you might get discouraged because you think there is no way that you could possibly memorize the names of 4,500 minerals! Here's the good news – you really only need to know about **12 minerals** to have fun with rocks. These very special minerals are known as **the rock-forming minerals**. These rock-forming minerals make up the majority of the rocks and especially the ones that you come across every day. Many rocks do contain traces of a lot of other minerals, but just knowing 12 minerals can help you to identify most of the rocks. And to make it easier yet, there are six dark-colored rock-forming minerals and six light-colored rock-forming minerals. The twelve rock-forming minerals are as follows, and can be seen in the following pictures:

- Dark-colored rock-forming minerals: amphibole, pyroxene, calcium feldspar, biotite mica, iron (usually found in ores, like magnetite), and olivine.
- Light-colored rock-forming minerals: quartz, muscovite mica, potassium feldspar, sodium feldspar, jasper, and calcite.

Activity: Take out the rock-forming minerals from your kit and study them as you look at the charts below. Be sure to notice their chemical or atomic formulas. Compare the light- and dark-colored rock-forming minerals. Notice the elements that are common to the dark-colored rock-forming minerals, and those that are common to the light-colored rock-forming minerals. Record those in your notebook.

The Six Dark-Colored Rock-forming Minerals

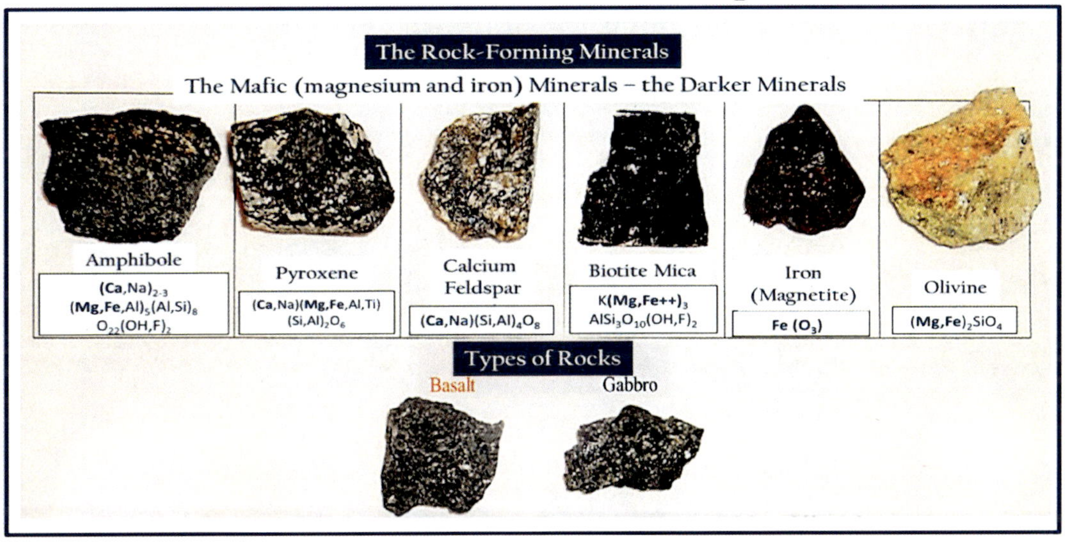

The Six Light-Colored Rock-forming Minerals:

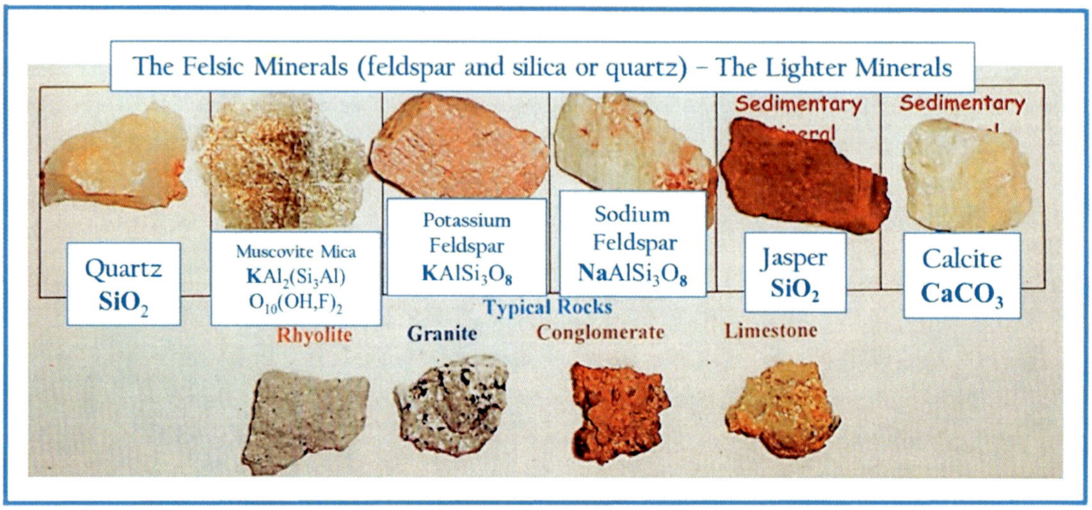

Notice that the dark-colored rock-forming minerals will be high in the elements **magnesium** (Mg) and **iron (Fe)**. That is why we call these minerals the **MAFIC** minerals, from the word *magnesium*, and the Latin word for iron, *ferric*. The light-colored rock-forming minerals will be high in the **elements potassium (K), sodium (Na), silicon (Si), and oxygen (O).** That is why we call these minerals, the **FELSIC** minerals, from *feldspar* (the lighter-colored feldspars), and *silica* (or quartz – which is silicon + oxygen). Let's put this knowledge to work.

Activity: Which rocks are made of the dark-colored rock-forming minerals and which are made of the light-colored rock-forming minerals? Put a (D) under those rocks that are made with the dark-colored rock-forming minerals and an (L) under those rocks that are made of the light-colored rock-forming minerals. (Answers to this exercise are found at the end of the activity.)

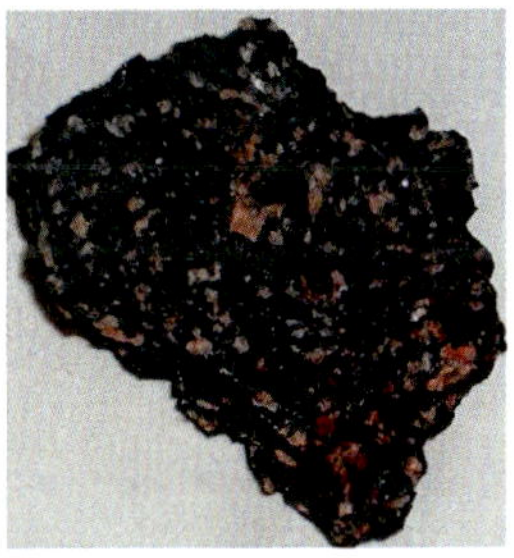

Answers: 1st Row: L, L, D
Answers: 2nd Row: D, L, L
Answers: 3rd Row: L, D, D,
Answers: 4th Row: L, D, D
Answers: 5th Row: L, D, D
Answers: 6th Row: D, L, L

Activity: Memorize the eight most abundant elements in the earth's crust and the 12 rock-forming minerals.

Lesson Four – What the Minerals Look Like in the Rocks

Word bank: *thin sections, petrography, polarizing microscope*

Geologists identify rocks by the minerals they contain. The various rock types are usually categorized as plutonic, volcanic (or igneous), metamorphic, and sedimentary. [2] All the rock types are made up predominantly of the rock-forming minerals. Some rocks have minerals that may be unique to that particular rock, and are not one of the twelve rock-forming minerals. These are called incidental minerals. For example, the mineral garnet is found in some metamorphic rocks, but not in all metamorphic rocks.

(Left) Schist with garnets; (Right) Schist without garnets

We have already seen that the darker colored minerals form the darker colored rocks. The lighter colored minerals form the lighter colored rocks.

Now, there is another way geologists identify rocks. They do this by noticing the percentage of the minerals in the rocks. How do they do this? There are two systems geologists use.

1. The precise way to do this is through what are called *thin sections*. The study of thin sections is called petrography (literally, *rock writing.)* A thin section is a laboratory preparation of a rock, mineral, soil, pottery, bone, or even metal sample for use with a polarizing petrographic microscope. The word *polarize* in this context means to divide light into various components. So, a polarizing microscope is a special microscope that organizes color patterns of various minerals within the rock's thin section.

---

[2] For further study on rock types and identification, see the book **Rock Identification Made Easy** at NorthwestRockAndFossil.com.

A polarizing microscope

To make a thin section, a thin sliver of rock is cut from the sample with a diamond saw and ground flat. It is ground so thin that it can be mounted onto a glass microscope slide. The polarizing microscope reveals color patterns of the various minerals through a special filter.

Examples of polarized thin sections of granodiorite and gabbro. The different colors signify different minerals.

These color patterns have been catalogued for identification in what is called the Michel-Levy color chart. This is a very expensive and time-consuming process, but it is fascinating to study.

You can also purchase pre-made thin sections, and they are also very expensive. Or you can purchase books showing photographs of the various thin sections that have been made and studied. Ok, this is not "Mineral Identification Made Easy!" Let's look at another process that might help to make things a little simpler.

2. Another way that geologists use is what I call, "The guess and by golly method." It is really an experienced guess as to the percentages of various minerals in a rock. Geologists use approximate percentages of mineral content to determine a rock's identification by the amount of a particular mineral it contains. (See chart below.) This is certainly less expensive and just requires practice. It is obviously not as accurate, but can give you a fairly good estimate of the rock type you are studying by noticing the approximate percentage of mineral content.

Visual chart showing the approximate mineral percentages in rocks for rock identification

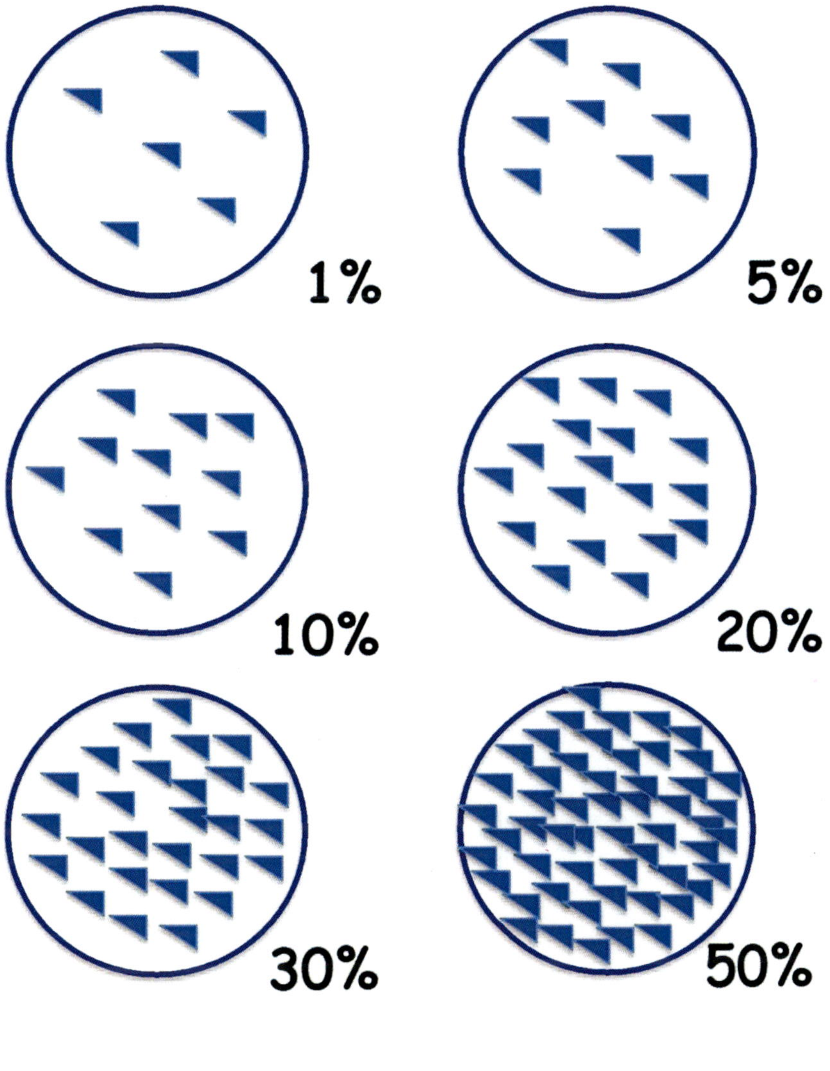

Granite with about 10-20% darker minerals and the remainder in lighter-colored minerals, potassium feldspar and quartz; this qualifies it as granite.

Percentages of dark and light minerals help define and categorize particularly the plutonic and volcanic rock types.

In the previous lesson, we talked about LIGHT and DARK minerals that we use to categorize various rocks. But what if the pattern you see does not fit totally one or the other? What then? Let me give you an example.

Granodiorite

The previous picture of granodiorite is a plutonic rock type. That means you can easily see the minerals in the rock. But what do you say when asked the question, is this rock light or dark? About 50% of people would say dark.

And about 50% of people would say light. We call these kinds of rocks, *intermediate*. In other words, they are between light and dark.

Another name for granodiorite is called a *salt and pepper* rock, meaning that it has an even mix of light and dark minerals. If you look at the chart above, you will notice that it would fit a 50% mix.

Another example of a *salt and pepper* rock is diorite. It is also a plutonic rock with a 50/50 mixture of light and dark minerals. The main difference between diorite and granodiorite is that granodiorite has more quartz.

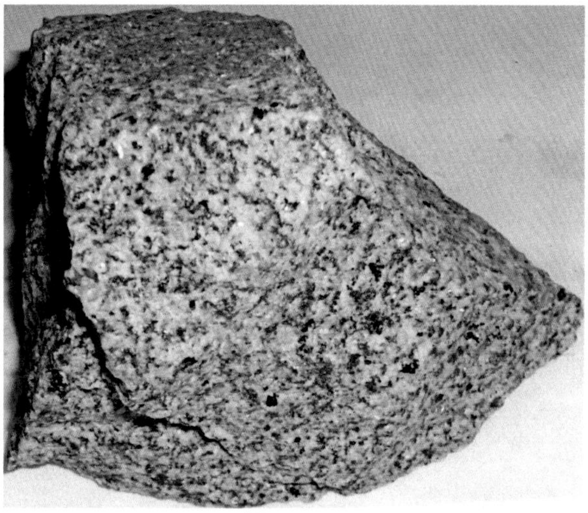

Diorite

Geologists use another special chart that shows the percentages of minerals in rocks based on thin sections. This particular chart shows the approximate percentages by weight and volume of different minerals in volcanic and plutonic rocks. The number down the left side represents the percentage by volume of the minerals that are given within the box. Across the bottom, the percentage of quartz (SiO_2) by weight is given. Although there is lots of technical information on this graph, I don't want you to get bogged down in that. The rocks that are on the left side of the chart (rhyolite and granite) are considered light-colored rocks, with higher percentages of light-colored rock-forming minerals. The rocks on the right side of the chart (komatiite, peridotite, basalt, and gabbro) are those that are considered dark-colored rocks, made from primarily dark-colored rock-forming minerals. For our purposes, it is important to note that with increasingly darker rocks, the mineral percentage of quartz, a light-colored rock-forming mineral, decreases (the numbers across the bottom of the graph).

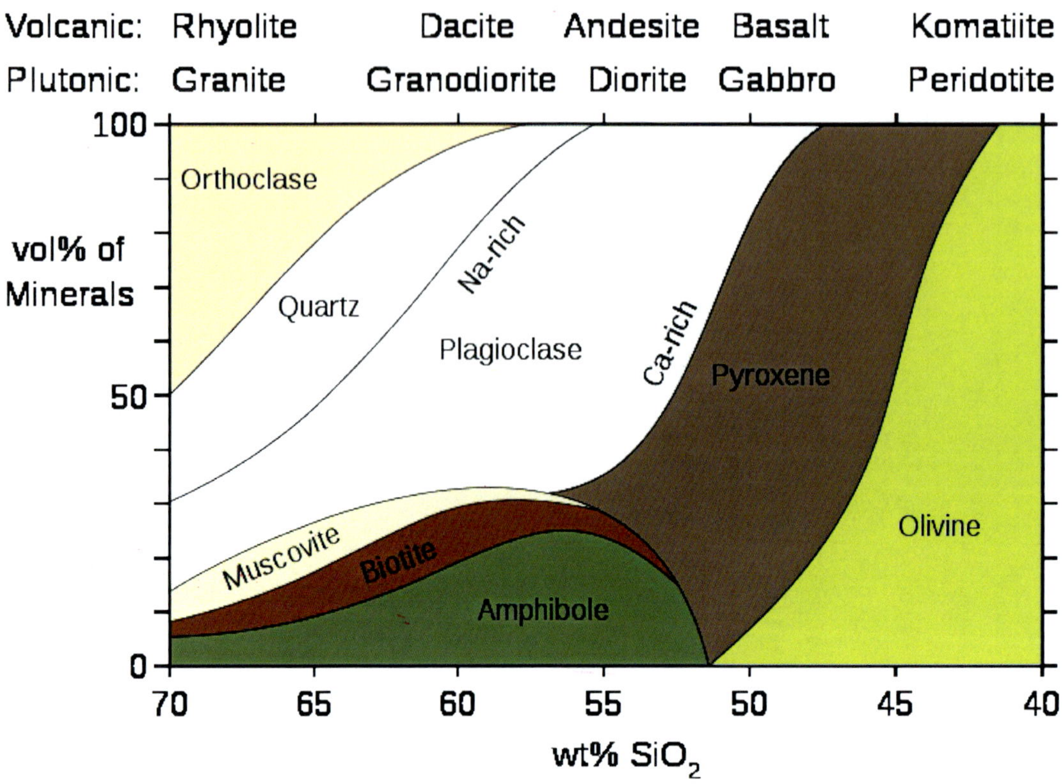

For example, you will notice that gabbro (a plutonic rock) has a greater percentage of calcium feldspar (Ca-rich plagioclase) than many other rocks. Notice, also, that gabbro has a low percentage of quartz. Because it has predominantly more Ca-rich plagioclase, gabbro is considered a dark-colored rock. The remaining rocks (dacite, granodiorite, andesite, and diorite) are what are called intermediate rocks, as they contain a very even mix of both dark- and light-colored minerals.

Gabbro: rich in calcium feldspar, but low in quartz

Gabbro – although it has some lighter colored plagioclase feldspar in its mix, it is about 80% darker minerals like calcium feldspar, pyroxene, biotite, and amphibole.

When you look at an advanced rock identification book, you will be told that a particular rock has a particular percentage mix of certain minerals. You may wonder how they got those percentages? They either got it using *thin sections* as I mentioned before, or through the *by guess and by golly* method.

If you combine the above two charts you can become proficient at identifying rocks based on their mineral content percentages.

Let's take our charts and apply them to some volcanic rocks.

Rhyolite block from Yellowstone Park

According to our previous chart, this sample of rhyolite contains an abundance of the salmon colored mineral called orthoclase. And it contains an abundance of quartz. These two things taken together make it a lighter-colored rock.

Dacite

This sample of dacite borders on being a light-colored rock but is still considered an intermediate volcanic rock. Notice too that it is missing the color that the rhyolite had. If we look at our chart above, we will notice that dacite has quartz and sodium feldspar (plagioclase) right between the rhyolite and the andesite.

Andesite is an intermediate volcanic rock

On the other side of the chart above are rocks that have high amounts of the darker minerals like pyroxene and olivine. And basalt certainly qualifies as the rock that would fit the profile – dark and with very little quartz.

Basalt

With some experience at using these two charts you can become quite proficient at recognizing a rock simply by the estimated percentage of mineral content.

What about sedimentary rocks? Sedimentary rocks are those that are made of mud or chemicals. Most of the sedimentary rocks are made up of the lighter-colored rock-forming minerals, **calcite** and **quartz**, but the predominate mineral is calcite. Again, it is the combination of various elements that give these rocks their various shades of color.

This reddish piece of sandstone is made of tiny quartz crystals. It is colored by iron to give it a very pretty orange to red color. The travertine on the right is a type of limestone with bands colored by the element iron.

Coal is a sedimentary rock. But its high content of the element carbon makes it black, due mostly to fossil plants that have been altered from the heat and pressure of overlying sediments. Plants are high in carbon. The sample of limestone on the right is a reddish color due to the presence of the element iron.

Limestone varieties formed from the light-colored rock-forming mineral, calcite

What about metamorphic rocks? Although it has never been seen forming, metamorphic rock is thought to be the result of an original rock that has been altered by heat and pressure.

Gneiss - The light-colored mineral is potassium feldspar and the dark-colored mineral that forms the bands is pyroxene.

Here are two samples of gneiss with the same mix of light and dark-colored minerals, with the addition of lots of flakes of the mineral mica, reminiscent of the rock granite - more of a *salt and pepper* mix of minerals.

Activity: Look up, either in this book, another book, or online, pictures of different minerals in their crystalline shape. In your notebook, draw and color pictures of at least three of them. Be sure to try to capture the representative shape of the crystal.

Lesson Five – The Mineral Families

Word bank: *Native element*

All minerals can be grouped into families of minerals. All families of minerals are identified by specific chemical arrangements of certain elements that all members of that family share.

The mineral families are:

1. The **Silicate** family
2. The **Oxide** family
3. The **Phosphate** family
4. The **Sulphate** family
5. The **Sulfide** family
6. The **Halide** family
7. The **Carbonate** family
8. The **Native Element** family

Some people also suggest a ninth family called the **mineraloid family**. The term is used for those substances that do not fit neatly into one of these eight classes. Even though the chemical properties could fit into one of the members above, mineraloids generally have *no identifiable crystal structure*. The mineraloids include:

- Amber, non-crystalline structure, organic (fossil tree sap)
- Ebonite, vulcanized natural or synthetic rubber (organic); lacks a crystalline structure
- Jet, non-crystalline nature, organic (very compact coal)
- Lechatelierite, nearly pure silica glass (volcanically produced)
- Limonite, a mixture of oxides and hydroxides of iron (usually considered an ore of iron)
- Mercury, liquid, interesting stuff but dangerous!
- Obsidian, volcanic glass - non-crystalline structure, quartz and iron mixture. Obsidian is also classified as a volcanic rock, a type of rhyolite lava that has been colored by iron.
- Opal, non-crystalline silicon dioxide, and a mix of other minerals. Opal is a silicate mineral in chemical composition, but is believed to have been altered by very hot water. It is extremely soft and doesn't behave like quartz, which is a very hard mineral.
- Pearl, organically produced carbonate

- Petroleum, liquid, organic
- Pyrobitumen, amorphous fossilized petroleum (noncrystalline, organic)
- Tektites, meteoritic silica glass with other elements

Now let's take a closer look at each of the eight families of minerals and see why they are grouped into these families. This is where a little knowledge of chemistry would come in handy.

1. **The Silicate family** is the largest of the mineral families. It contains the most minerals. The chemical formula of a silicate mineral contains silicon **(Si)** and oxygen **(O)** grouped together. The silicate minerals have a lot of quartz, identified by the quartz formula, **SiO_2.** Let's look at two silicate minerals and see if we can spot the pattern.
   - **Willemite** – Zn_2SiO_4. This mineral has the elements silicon and oxygen, together. So whenever you see the **SiO** combination, you have a silicate mineral.

Willemite is also fluorescent; look for the green color in the above picture.

   - **Kyanite** – The chemical formula for kyanite is Al_2**SiO_5**. Since it has silicon and oxygen together, it is considered to be a part of the silicate family.

Kyanite

2. **The Oxide family** always has the oxygen **(O)** element accompanied by other elements. So whenever you see a formula where you have the O along with other elements, you have an oxide.

- **Cuprite** (copper oxide) is an example. Its chemical formula is Cu_2O – copper combined with oxygen.

A cuprite crystal – copper oxide

- **Magnetite** (iron oxide) Its chemical formula is Fe_3O_4: iron **(Fe)** combined with oxygen. Magnetite is often magnetic.

A magnetite crystal

3. **The Phosphate family** always has the symbols, **PO** (phosphorus and oxygen together), accompanied by other elements. If you see this combination, you can group that mineral with the phosphate family of minerals.
   - **Turquoise** (a source of copper, Cu) is a phosphate mineral with the chemical formula, $CuAl_6(\mathbf{PO_4})_4(OH)_8 \cdot 4H_2O$: it is prized for jewelry.

Bright blue green turquoise

- **Apatite**, another example of a phosphate mineral, is also used in the Mohs Hardness Scale with the hardness of 5. We will talk about the Mohs minerals later. It's chemical formula is $Ca_5(PO_4)_3(F,Cl,OH)$.

Green apatite: it is also common in blue cryptocrystalline masses.

4. **The Sulphate family** always has the symbol SO (sulphur and oxygen together) accompanied by other elements.
   - **Barite** – the word means *heavy spar*. And it is heavy! It is often found with metals like galena or another lead ore, called cerussite. It also has several different forms in which it can appear. The chemical formula for barite is $BaSO_4$ (barium sulfate).

Barite

- **Epsomite** – from which we get Epsom salts for medicinal purposes. Its chemical formula is $MgSO_47H_2O$.

Epsomite

5. **The Sulfide family** always has the element sulfur **(S)** accompanied with other elements.
   - **Sphalerite** comes in many colors, including yellow, brown, and black. It is quite heavy. Its chemical formula is ZnS (zinc sulfide).

Sphalerite crystals

   - **Pyrite is** otherwise known as *fool's gold* because it fooled a lot of the novice miners in the 1800s who had travelled west to seek their fortune in the gold fields. Its chemical formula is FeS_2 (iron sulfide).

Pyrite crystals (also known as iron pyrite); it is an ore of iron.

6. **The Halide family** always has one of the following elements fluorine **(Fl)**, chlorine **(Cl)**, bromine **(Br)**, or iodine **(I)** accompanied with other elements.

• **Halite is** otherwise known as *rock salt*. Its chemical formula is Na**Cl**, sodium chloride.

Halite - salt crystals

• **Fluorite** – chemical symbol is CaF_2, calcium fluoride.

Natural fluorite

• **Bromargyrite** – chemical symbol is Ag**Br**, silver bromide.

Bromargyrite

- **Miersite** – chemical symbol is (Ag,Cu)**I**, silver copper iodide. Not a common mineral but one that includes the element **iodine**, so it is in the halide family.

Miersite – an ore of copper

7. **The Carbonate family** – always consists of the elements carbon and oxygen **(CO)** together) accompanied with other elements.
    - **Calcite** – chemical symbol is $CaCO_3$ (calcium carbonate).

One of the many shapes in which calcite can appear. Because of the many different crystal patterns it can take on, the way to tell if you have calcite is to test it with an acid. It will fizz.

- **Cerussite** or lead ore. Its chemical symbol is $PbCO_3$ (lead carbonate). You might never expect to see lead ore looking like this! It looks like quartz but because of its unique chemical formula, it is an ore of lead, which is indicated by Pb (lead).

A cerussite crystal

8. **The Native Element family** – consists of just one element by itself. Some examples include the following:
   - Copper – Cu
   - Gold – Au
   - Silver – Ag

Native copper

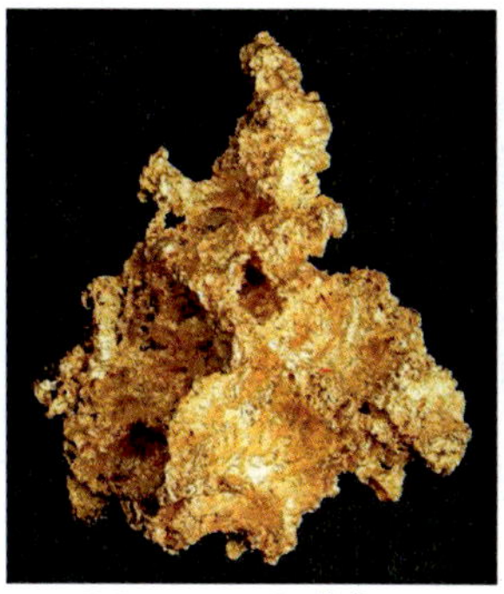

Native Gold

Native Silver

Activity: Using the worksheet provided below, look up at least five additional minerals from each family and record their chemical formulas and mineral families. I would suggest that you start by looking up the mineral family name first.

Worksheet for Mineral Families

Mineral Name	Chemical Formula	Mineral Family

Lesson Six – The Feldspar Family of Minerals

Word bank: alkali, feldspar

Even though the feldspars are all silicate minerals and, therefore, would belong to the silicate family of minerals, it is such a large group that we can treat it as a subset of the silicate family. Why does feldspar have its own subset of minerals? Together the family of feldspars makes up about 60% of the earth's crust. You will, therefore, run into it a lot. The feldspars are in large part why rocks are the color they are.

We get our word feldspar after a long trip around the block, so to speak. The word feldspar comes from a German word, *feldspat*, a compound of the words *feld*, field, and *spat*, meaning, a rock that does not contain ore. The change from *spat* to -spar was influenced by the English word *spar*, a synonym for mineral. *Feldspathic* refers to materials that contain feldspar. Now, that doesn't help you identify it any easier, but just introduces the mineral. It is not always easy to identify, either.

So, what is feldspar? Feldspar is a member of the Silicate family of minerals. Of the feldspars, there are three that are more common that you find in rocks. They are orthoclase feldspar (potassium feldspar), albite (sodium feldspar), and labradorite (calcium feldspar). But each feldspar has its own chemical characteristic. All feldspars revolve around quartz, sodium, potassium and calcium. Let's examine how the feldspars are organized.

There are two groups of feldspars:

The **alkali** feldspars all contain **potassium** and quartz, and are typically light-colored.

- **Orthoclase** feldspar or **K**-spar with the chemical formula, $KAlSi_3O_8$ (**potassium**, aluminum, silicon and oxygen)
- **Sanidine** feldspar with the chemical formula, $(K,Na)AlSi_3O_8$ (**potassium**, **sodium**, aluminum, silicon and oxygen
- **Microcline** feldspar with the chemical formula, $KAlSi_3O_8$ (**potassium**, aluminum, silicon and oxygen; similar to orthoclase except in crystal shape
- **Anorthoclase** feldspar with the chemical formula, $(Na,K)AlSi_3O_8$ (**sodium**, **potassium**, aluminum, silicon and oxygen)

Now, since the goal of this book is to make mineral identification easy, we are not going to get into all the fine points about the different feldspars. Suffice it to say that many rocks are salmon, pink or white in color due to the **alkali** feldspars. Remember, the alkali feldspars all contain potassium which is light-colored.

The pinkish color in the rhyolite breccia and white color in granite are due to the alkali feldspars.

The colors in the pegmatite on the left and in the rhyolite on the right are all due to presence of alkali feldspars.

The pinkish color in these two samples of rhyolite are due to the alkali feldspars.

The second group of feldspars is called **plagioclase** feldspar. These feldspars all have different degrees of either **sodium** or **calcium**. And the rule of thumb is that the more calcium it has, the darker the feldspar and the rock that contains it.

- **Albite** feldspar with the chemical formula, $NaAlSi_3O_8$ (**sodium**, aluminum, silicon and oxygen)
- **Oligoclase** feldspar (also known as *sunstone*), with the chemical formula, $(Na,Ca)(Al,Si)AlSi_2O_8$ (**sodium**, **calcium**, aluminum, silicon and oxygen)
- **Andesine** feldspar with the chemical formula, $NaAlSi_3O_8$—$CaAl_2Si_2O_8$ ranging from **sodium** to **calcium**; (**sodium**, aluminum, silicon, oxygen to **calcium**, aluminum, silicon and oxygen)
- **Labradorite** feldspar with the chemical formula, $(Ca,Na)Al(Al,Si)Si_2O_8$ (**calcium**, **sodium**, aluminum, silicon and oxygen)
- **Bytownite** feldspar with the chemical formula, $(NaSi,CaAl)AlSi_2O_8$ (**sodium**, silicon, **calcium**, aluminum, silicon, and oxygen)
- **Anorthite** feldspar with the chemical formula, $CaAl_2Si_2O_8$ (**Calcium**, aluminum, silicon and oxygen)

(Left) The rock, anorthosite is almost completely made of the feldspar anorthite. (Middle) This lava is a dark-colored rock; if you look closely you can see the white sodium feldspar minerals as needles. (Right) Dacite lava owes its light color to both quartz and albite or andesine feldspar.

Here is a section of the basalt lava enlarged to show the white needle-like sodium feldspar crystals.

In this sample of diorite, a plutonic rock, the white feldspar crystals are sodium feldspar.

These two samples of gabbro have a lot of calcium feldspar (bytownite).

The blue indigo gabbro on the left has a lot of calcium, as well as the samples of gabbro on the right.

When feldspar crystals are isolated, they generally appear either as needle-shaped whitish crystals or blocky whitish crystals as in the examples of andesite below.

Blocky whitish crystals of feldspar easily identify the two rocks above as andesite.

Feldspar plays a huge role in the rocks of the earth's crust in primarily determining their color. Other minerals do also play a role in color, such as iron, but predominantly it is the feldspars.

Basalt ribbon bomb: the reddish color is due to oxidation of the iron content of basalt.

Activity: Take out the feldspar collection from your kit and examine each one. See if you can spot the differences among them. In your notebook, draw pictures of the different feldspars and illustrate their differences.

Lesson Seven – Identifying Minerals

Word bank: *color (as applied to minerals), density, luster, Mohs hardness scale, streak test*

There is generally not one way or process used to identify a mineral. Some minerals can be easily identified because of experience. Most minerals take time and the identification of several properties of each mineral. Below is a list of the different ways minerals can be identified. These are called, **properties** of minerals.

1. **The Mohs Hardness Test Kit** is designed to help you narrow down your possibilities in identifying your mineral. It is a combination of simple tests that will help you eliminate possibilities. PLEASE NOTE: THESE TESTS WILL NOT WORK FOR TESTING THE HARDNESS OF ROCKS. Many people make that mistake and end up frustrated because they cannot make the Mohs Scale work for them. It is not meant for rocks, but for minerals. Why is this? It is because rocks are made up of one or more minerals. In the case of the rock granite, the minerals quartz, mica and feldspar all have a different hardness. Depending on where you test the granite, you could end up with different answers.

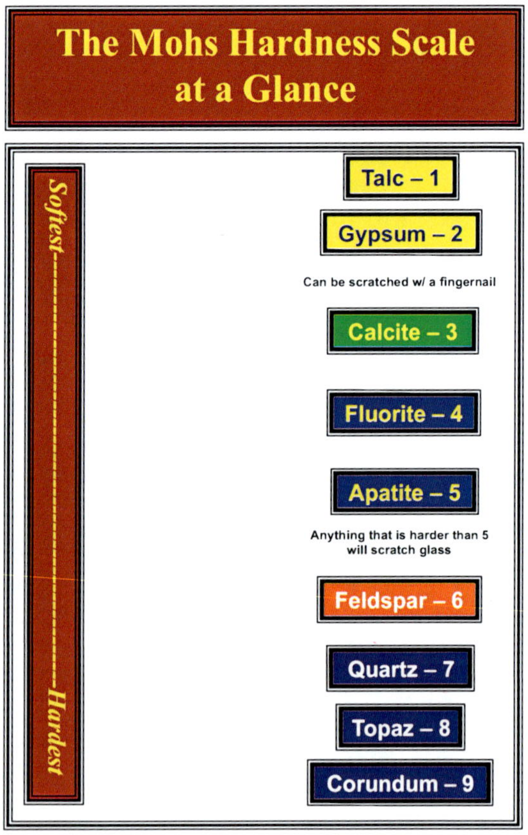

In 1828 Friedrich Mohs came up with a scale from one to ten, which gave a relative hardness rating to minerals. Talc was the softest mineral at a relative hardness of 1, and diamond was the hardest mineral at 10. He also developed a streak test for each mineral.

a) **The streak plate** – The purpose of the streak plate is to reveal the color that a mineral leaves. When it is scratched across a non-glazed, porcelain plate or tile, every mineral leaves a streak. This can help you narrow down some of your choices as to what the mineral is. Normally you will have two streak plates, a white and a black streak plate. You will use the plate that is most likely to show the color of your mineral. This test is only a relative test. It may help you get in the ballpark.
b) **The glass scratch test** – the hardness of glass is approximately a 5 on the Mohs Hardness scale of 1-10. So the idea here is that minerals softer than five will not scratch the glass plate. Minerals that are harder than glass will scratch glass, which means that they will have a hardness of between 5 and 10. But again, this is only a relative test. There are over 4,500 minerals and many of them show the same scratch test result.

The standard Mohs minerals and what they look like:

Talc (1) – a very soft mineral that can be scratched with the fingernail; it has a waxy feel to it.

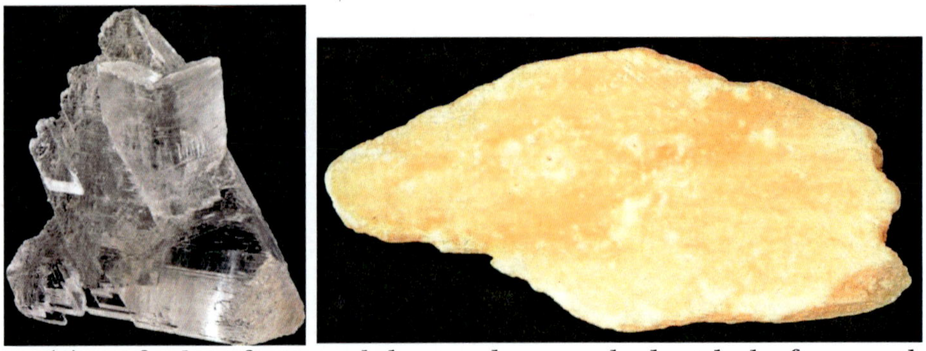

Gypsum (2) – a fairly soft mineral that can be scratched with the fingernail; it has many shapes and colors it can assume.

Calcite (3) – a fairly soft mineral that can assume many different shapes and colors. One sure test for calcite is to drop a small amount of a strong acid on it. It will fizz if it is calcite.

Fluorite (4) – occurring as greenish masses, it is readily identifiable. The dipyramidal (or octahedral shape, two pyramids attached at the base, having eight sides) is also a good indicator. One other test for fluorite that is not always consistent is that it will generally glow a deep blue/purple color under long or shortwave ultraviolet light.

Apatite (5) – Blue or green mineral that can be cryptocrystalline or occurring as hexagonal crystals

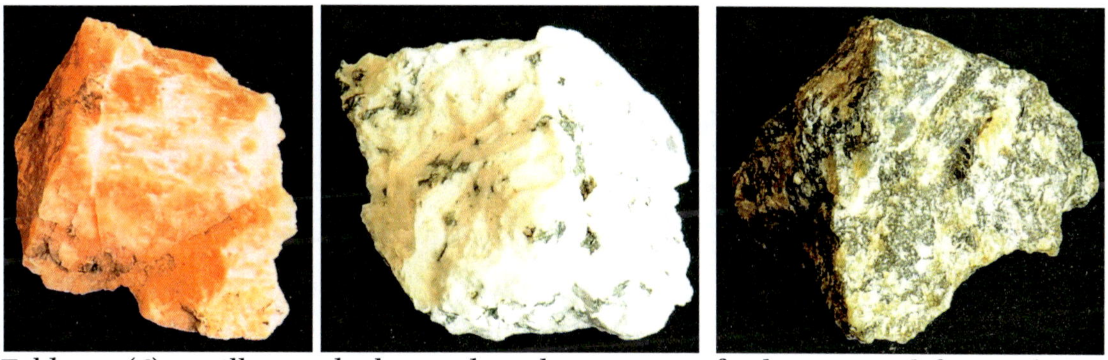

Feldspar (6) – will scratch glass and can be a variety of colors. From left to right are potassium feldspar, sodium feldspar and calcium feldspar. Generally, the feldspar minerals are blocky in their crystalline form.

Orthoclase feldspar crystal Albite feldspar crystal Anorthite feldspar crystal

Quartz (7) – as you can see, quartz can come in all sorts of colors. But its crystal pattern is always six-sided.

Topaz (8) – a very hard mineral and does have a hexagonal crystal pattern.

Corundum (9) - which occurs as ruby (and sapphire when blue): a very hard mineral that is used in abrasives. It is almost as hard as a diamond. The crystals occur as six-sided cylinders or barrels.

2. **Color** – although not a good indicator, color may again help you narrow down your options for identification. Keep in mind, though, many minerals share the same color and any one mineral may have several colors.
3. **Magnetism** – is particularly helpful when sorting through the metals and sulfides. Unless you have a magnetite crystal and a hematite crystal, which look entirely different, magnetism may help you tell the difference. But keep in mind that this is only a relative test. Some hematite is actually magnetic!
4. **Density** – generally, the heavier a mineral is, the denser it is. The density of calcite is 2.71. The density of barite is 4.48, which is very heavy. And unless you have the sophisticated equipment to make yourself an apparatus for measuring density, I would suggest that you get a book like *The Smithsonian Guide to Minerals* or some other mineral book that would have all the pertinent density information recorded in it for just about every mineral that is known. Another way to learn about the density of a mineral is sometimes by its name. For example, if you are trying to decide between two minerals that have a similar appearance, like barite and calcite, it helps to know that the name barite means heavy spar. (The term **spar** is a generic term used by geologists to refer to any non-metallic mineral with a glassy luster that breaks along distinct flat surfaces). This could be an indication as to which is barite.

Classic orange and white barite crystal patterns, but remember, it can also appear as blocky spar, very like calcite spar.

Beryl would be another example. It is a six-sided barrel shaped mineral. So is corundum. Beryl's density is 2.76. Corundum's density is 4.05, a very heavy mineral. Again, the density would be the deciding factor as to which is which. Both beryl and corundum can also be blue, confusing the identification between blue beryl and sapphire. Density again will be the deciding factor.

Beryl and corundum are both six-sided crystals, but one is denser than the other.

Density can be the deciding factor for many minerals. It is a very useful property to use in your identification of minerals.

5. **Crystal shape or pattern (also known as its *crystal habit*)** – this can be a good identifier, but only if other properties are considered along with it. Many minerals share the same shape. But it can be a good start. The six-sided clear mineral is a dead giveaway for quartz. So is the purple six-sided amethyst – quartz. But this is not always going

to be the case in mineral identification.

Amethyst

6. **Luster** – is the way light interacts with the surface of a crystal, rock, or mineral. The word traces its origins back to the Latin word *lux*, meaning light, and generally implies radiance, gloss, or brilliance. A metallic luster would accurately describe a metal or metallic ore.

   But there are also other terms to become familiar with when describing a mineral's luster such as *earthy*, *metallic*, *greasy*, *silky*, and *vitreous*. Vitreous is from the Latin word for glass, *vitrum*, and refers to a glassy luster. Luster is a valuable identifier for minerals, and it is fairly accurate.

Both gold and pyrite have a metallic luster.

Opal has a greasy luster to it

A variety of gypsum called satin spar has a silky luster to it.

(Left) Kaolin (a type of hard clay) has an earthy luster to it.
(Right) Limonite (a source for iron) also has an earthy luster to it.

Both biotite and muscovite mica have a pearly luster to them.

This example of quartz has a vitreous or glassy luster to it

At times I just cannot identify a mineral. Even after consulting all the books I can't be sure. At that point I will make sure I catalogue it with an identifying number as to where I found it. In time I will usually run across another example of it somewhere else or read a paper about the same mineral. It is a journey! The most important thing – RECORD WHERE YOU FOUND IT!

Activity: Using the worksheet provided below, see if you can organize your minerals according to their properties. Your process would include:

1. Using the glass plate, gently attempt to scratch the glass plate. If it does not scratch the plate, then record ‹5 on your chart. If the mineral scratches the glass plate, then record ›5. To get more specific, you can consult a more detailed minerology resource for its specific hardness.
2. Take out the porcelain plates from your kit and using the selected mineral, carefully scratch the surface of the plates. Record the color that you see.
3. Compare the mineral with the minerals from your kit that show crystalline habits. Choose the one that is most like it and record that observation. If your sample is cryptocrystalline, make that note on your chart.
4. Using an internet source, look up the name of your mineral and record its density.
5. Does your mineral have a metallic luster, a shiny, glassy luster, or a dull luster? Record your answer.
6. Record the color of your mineral.

7. Using the magnet from your Mohs hardness test kit, test your mineral to see if it displays any magnetic properties. Record your answer.

Worksheet for Mineral Properties

Mineral	Hardness	Streak Color	Crystal Habit	Density	Luster	Color	Magnetism

Lesson Eight – Building Your Mineral Collection

Most people pick up so many rocks and minerals that they easily lose track of them and then forget where they got them. They treat their collection the same way we gather our tax receipts for the accountant – in a shoebox! I have had more kids show me their rock collections in a shoebox. And when I ask them where they got such and such a rock or mineral, they cannot remember.

Being able to remember where you found your specimens, even if they are purchased, is part of the story of your interest. And the area where you found your rock or mineral can actually help you identify the rock and mineral.

There is an easy and inexpensive way to start building your collection. Here's how you do it.

1. **Collect your old egg cartons!** The spaces are usually large enough to hold most samples. The larger samples of rocks and minerals are called *display* samples and they don't belong in egg cartons but on a shelf where everyone can admire them.

Egg cartons are an inexpensive way to utilize something you need anyway

2. Most kids have access to using PowerPoint on their computers. Design an identification sheet that you can put with your samples. Be sure to come up with a numbering system and store that in a safe place. You can make this as colorful as you like. I use a color code system – dark blue for plutonic rocks, light blue for minerals, red for metamorphic rocks, brown for sedimentary rocks, and orange for volcanic rocks. But you can use any scheme you want. You can even

organize your minerals according to the mineral families labeling each mineral with its chemical composition.

3. **Paint or color your egg carton.** Again, your egg cartons could be color-coded. That way you can easily pick out which collection you want to study or show someone. But if your rocks and minerals are all jumbled together in a shoebox, well that's just not very organized or attractive.

4. When you start amassing a huge collection like the one I have, it takes a lot of attention to keep up with things. **So, I use a cabinet with various sized display boxes to organize my collection.** Display cabinets can be extremely expensive. But occasionally you can find them at garage sales, estate sales. I found three ideal cabinets that way. And they were on sale! Anything that holds tools or hardware can also be used for rocks. Be creative!

To help you organize a nice rock and mineral collection I will give you four key ideas to remember:

1. **Record the location of your finds**
2. **Identify your finds**
3. **Organize your finds**
4. **Display your finds**

Activity: For practice:

1. Separate your minerals from the rocks, and put in these two groups.
2. Next, sort your minerals according to families.
3. See if you can sort your rocks according to their types.
4. Once you have done this, try displaying a few of your prized possessions by getting a few egg cartons and decorating them according to your mineral families and rock types. You may want to use a color code system.
5. Once you have organized your collection into the various compartments of each egg carton, create an attractive label for each of your displays.

Lesson Nine – Working with Fluorescent Minerals

Word bank: *fluorescence, hydrothermal, longwave, midwave, shortwave*

What is fluorescence? Briefly, **fluorescence is the ability of a rock or mineral to respond to ultraviolet light**. Although it can be a property of certain rocks and minerals, it is not a good indicator for mineral identification. It is not consistent, and the same mineral could have different results and even different colors. Nevertheless, it is a fun part of the mineral hobby. Everyone likes to see rocks and minerals glow in the dark when a special light is shown on them.

But why do some rocks and minerals glow and some don't, even of the same type of mineral or rock? And what is the origin of fluorescence? The answers to these questions are not known for sure!

However, one observation I have made is that many of the fluorescent rocks and minerals in my collection come from areas known for *hydrothermal alteration* – hot water alteration. Many of the famous fluorescent minerals come from volcanic areas like Arizona and Mexico. These areas are also known for their sedimentary rocks and valuable gems and metals. I do believe that fluorescence is related to hydrothermal alteration at the time of the Genesis Flood when conditions would have been just right for the altering of sedimentary rocks. At any rate, no one has ever seen a fluorescent rock or mineral form today.

There are three types of ultraviolet light that concern fluorescent rocks and minerals.

1. **Longwave** – *not dangerous to the eyes or skin*; It will show some things, but not very many. Another name for the longwave light is the *black light*. For rocks and minerals, however, you need a special filter that comes with most commercial rock and mineral lights. The filter helps keep the reflecting glare off of the rocks and minerals so that the true color comes out. Otherwise what you will always get back is the blue/purplish light from the bulb of the longwave light.
2. **Shortwave** – *dangerous to the eyes and skin.* Many minerals glow intensely under shortwave light. Shortwave light tends to give better results than the other light waves. (If you have children, please watch them very closely when handling this light. Not only

is it dangerous to skin and eyes, but these lights are rather expensive, and the replacement bulbs are not cheap!)

3. **Midwave** – *dangerous to the eyes and skin.* Shows up a few things that otherwise would not be visible under longwave or shortwave light.

A nice arrangement of fluorescent minerals

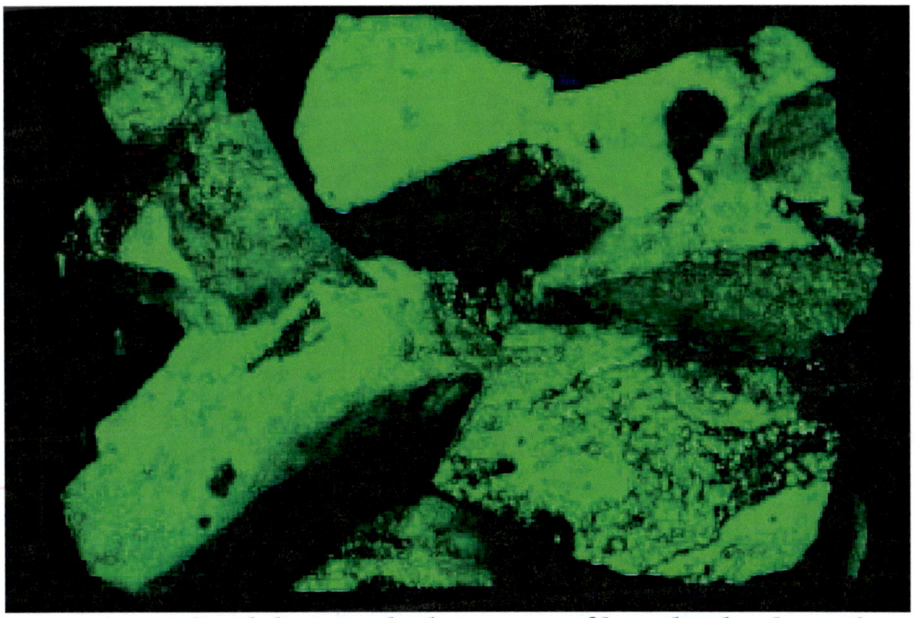

Fluorescent opalized dacite, which is a type of lava that has been altered somewhat. The fact that opal is involved tells me that the silica that the dacite once contained has undergone some sort of change to allow it to glow a brilliant green – perhaps through some kind of contact with very hot water that may have contained uranium.

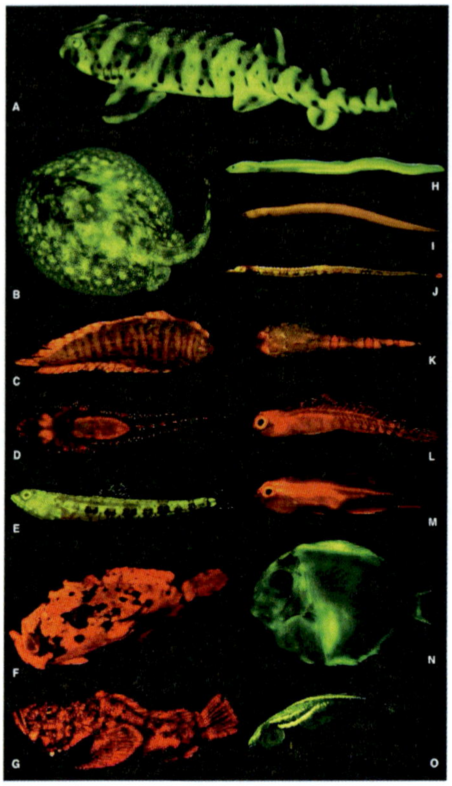

Fluorescent Fish

Fluorescence is really quite a mystery. In the pictures below these particular samples of calcite do different things under different light waves. But this is not consistent!

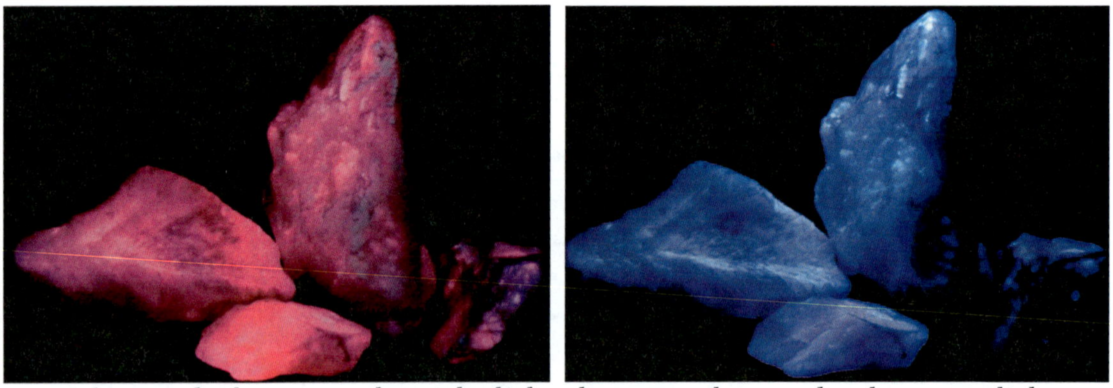
Calcite under longwave ultraviolet light; the same calcite under shortwave light

Fluorescent lights come in all sorts of wattages. A simple portable 4-watt light that operates on batteries will show up a lot of things, especially if it is a shortwave light. Don't waste your time picking up a black light from the hardware store. They do not have the necessary filter and black lights just do not show up the brilliant colors that shortwave lights do. **You must view the rocks and minerals in total darkness to get the proper effect.**

Fluorescence is not consistent. So, I like to keep a portable fluorescent light handy to test rocks and minerals on site. You will be surprised every time you test something. Things that you thought were fluorescent will not be and vice versa.[3]

Activity: If you have access to a fluorescent lamp, use the worksheets on the next few pages of your book to test and record the response of different materials to your lamp.

Fluorescence Worksheet (1)

Take out your ultraviolet light and begin to explore the different minerals and rocks you already have in your collection. Record your observations.

Fluorescence Worksheet (2)

Take out your ultraviolet light and begin to explore various materials within your home or garage, or in your backyard. Record your observations.

[3] Fluorescent kits with samples and a lamp are available at NorthwestRockAndFossil.com.

Fluorescence Worksheet (1)
Discovery Sheet for Minerals and Rocks

Mineral/Rock Name (if known)	Where did you find it?	Natural Color	Visible Crystals?	Is the mineral or rock phosphorescent?	Ultraviolet Color? (LW or SW)

Fluorescence Worksheet (2)
Discovery Sheet for Household Materials

Name of Item	Location	Natural Color	Visible Colors?	Is the material phosphorescent?	Ultraviolet Color? (LW or SW)

Lesson Ten – Gemstones

Word bank: gem (gemstone)

What is a gemstone? Very simply a gemstone is a piece of mineral which, when cut and polished, is used in making jewelry or other ornaments. But there are many other substances, not classed as minerals, that are made into jewelry and these are valued as gemstones too.

There is no universally recognized grading system for gemstones. The only organized standard in gemstones today is among the diamonds – the Gemological Institute of America or GIA developed in the early 1950s.

THE ULTIMATE GEMSTONE?

Did you know that diamonds are not rare? The diamond market is an artificial market that was created through the diamond mining company, DeBeers. Starting in the late 1800s, and on into the 1900s, they created a monopoly on the diamond market. Since then, by using clever marketing techniques, and limiting the quantity available, they have created the ultimate illusion that "...a diamond is a girl's best friend." Their revenue is around $6 billion per year.

Gemstones can include the following materials:

- **Lapis lazuli** is actually a metamorphic rock composed of **lazurite** (a very complicated mineral called a *tectosilicate* and a *feldspathoid* and a member of the sodalite group of silicate minerals. Notice its complex chemical formula – $Na,Ca)_8[(S,Cl,SO_4,OH)_2|(Al_6Si_6O_{24})]$ – *Wow!* It is complicated because it has characteristics of the bromine family (chlorine), the sulfate family (sulfur and oxygen together) and the silicate family (silicon and oxygen together). In spite of its complexity, people have used lapis lazuli, (lapis, for short) for thousands of years as jewelry. It has also been used as pigment for paint in the famous paintings of the Middle Ages primarily for the robes of angels and the Virgin Mary. In addition to the mineral lazurite, lapis is also made of calcite, sodalite (a blue silicate mineral), and pyrite, that gives lapis its gold colored streaks.

SAPPHIRES OR LAPIS LAZULI?

The word often translated into English as sapphire in the Old Testament was translated from the Latin Vulgate, which used the Latin word, *lapidus sapphiri*, which may in fact be a reference to Lapis lazuli. As late as the Middle Ages, lapis lazuli was often called sapphire (*sapphirus* in Latin, *sappir* in Hebrew), though it had little to do with the stone we know today as the *blue corundum* variety, *sapphire*. In his book on stones, the Greek scientist Theophrastus was probably describing the *sapphirus*, which is speckled with gold, as a description of lapis lazuli.

Natural Lapis lazuli; as adornment in the funeral mask of Tutankhamun

- **Pearls** are considered organic and so technically are not minerals. Remember, earlier in our study, we put them in the mineraloids family. A pearl is a hard object produced within the soft tissue of a living shelled mollusk. Just like the shell of a clam, a pearl is composed of the elements calcium, carbon and oxygen (the mineral calcite) in cryptocrystalline form, which has been deposited in concentric layers. Almost any shelled mollusk can produce some kind of pearl when an irritating microscopic object becomes trapped within its mantle folds, but the great majority of these pearls are not valued as gemstones. Pearls have also been used for thousands of years as prized jewelry.

Examples of pearls

- **Aventurine** is actually a metamorphic rock identified as *quartzite*, which is made of metamorphosed sandstone – quartz crystals that have been fused together to produce a very hard rock. It is called aventurine when it has a shimmering or glistening appearance called *aventurescence*. Aventurine (quartzite) can come in all sorts of colors. It is a very hard rock and polishes easily. It is one of my favorite stones to tumble.

(Left) Quartzite colored by chlorine
(Right) Quartzite when polished becomes a gemstone called aventurine.

Tumbled quartzite can be a beautiful stone for all kinds of jewelry.

- **Amber** is fossilized tree resin, which is organic, and, therefore, not considered to be a mineral. Remember that we put this in the mineraloids family. You might be most familiar with amber as that stuff with insects in it. But amber is found with all kinds of fossil critters in it from insects to water creatures! Amber is increasingly being made into jewelry, thanks in part to the Jurassic Park movies, as these have promoted an increased interest and awareness of the amazingly preserved fossils in it.

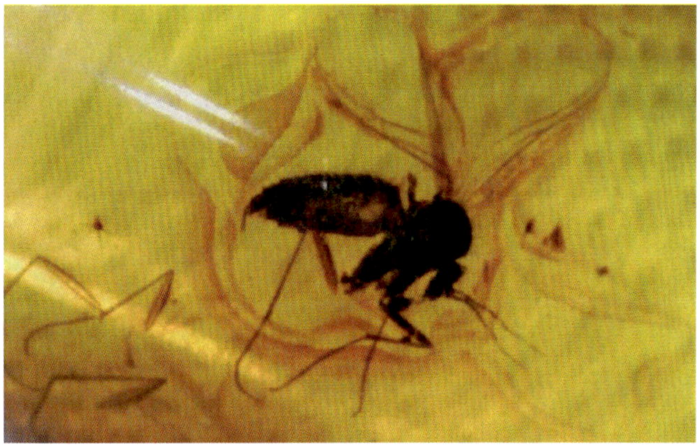

Fossil mosquito in amber

Jewelry pendants made out of amber

- **Jet** is a type of lignite coal. It is organic and, therefore, not considered to be a mineral. It is a mineraloid. But it does make for some very interesting jewelry. Jet is either black or dark brown, but may contain pyrite inclusions, which give it a brassy color and metallic luster. The phrase, *jet-black*, means as dark a black as possible and comes from this material.

Natural jet, or lignite coal

Mourning jewelry: jet brooch from the 19th century

- **Ammolite** is an opal-like organic gemstone found primarily along the eastern slopes of the Rocky Mountains of North America. It is made of the fossilized shells of ammonites, a type of fossil cephalopod, which in turn are composed primarily of the calcium carbonate mineral, aragonite. So, it is permineralized ammonite shell. The chemical composition of ammolite is variable, and in addition to aragonite, may include the minerals calcite, silica, pyrite, or other minerals. The shell itself may contain a number of trace elements, including the elements aluminum, barium, chromium, copper, iron, magnesium, manganese, strontium, titanium, and vanadium. These probably account for the fabulous colors of ammolite, which is fast becoming an expensive and highly sought-after gemstone. Because it is made of the shells of ammonites, ammolite is considered organic and, therefore, is not a mineral. Just how ammolite was formed is not known for sure. I suspect it has something to do with the saturated mineral-rich waters of the Genesis Flood and the heat generated by the abundance of volcanic activity associated with the same event.

(Left) Unprocessed ammolite
(Right) An iridescent ammonite

Ammolite jewelry

- **Opal** is hydrothermally altered silica with a chemical formula, $SiO_2 \cdot nH_2O$. But because it is amorphous, it is technically a mineraloid and not a mineral. It has, however, been highly prized as a gemstone for thousands of years.

Common opal

Opalized wood

Milky opal

Precious opal

Fire opal

Opal and diamond pendant

Mineral identification can be a lot of fun and a real challenge. Just remember that it is a combination of identifying factors that will generally give you the answer you are looking for.

Final Review for Mineral Identification Made Easy

All answers are easily found in the text.

1. What is the difference between a rock and a mineral? (Lesson One/Two)

2. From memory, reconstruct *The Building Blocks of Rocks* diagram. (Lesson Two)

3. What are the eight most abundant elements in the earth's crust? (Lesson Three)

4. Name the 12 rock-forming minerals. (Lesson Three)

5. Construct a diagram illustrating the mineral content of a rock that has 10% dark minerals, 25% dark minerals and 50% dark minerals. (See Lesson Four for an idea.)

6. Name the three types of feldspars most commonly seen in the rocks. (Lesson Six)

7. What is the predominant mineral in sedimentary rocks? (Lesson Four)

8. Give a brief description of a metamorphic rock. (Lesson Four)

9. Name the eight mineral families and their identifying elements. (lesson Five)

10. List five properties of minerals that could be used to identify them. (Lesson Seven)

11. Name the nine Mohs minerals and their hardness. (Lesson Seven)

12. What is the difference between streak and scratch? (Lesson Seven)

13. What is the difference between a mineral and a mineraloid? Give an example that illustrates each. (Lesson Five)

14. In your own words, give a brief description of a gemstone. (Lesson Ten)

15. What is ammolite? (Lesson Ten)

Index

Credits

Photos, charts, and images not credited are by Patrick Nurre or Vicki Nurre; samples in photos are from the author's collection. All photos by Heidi Noggle are used by permission, and samples are from the author's collection.

Title page: Quartz crystal, by Didier Descouens - Own work, CC BY-SA 3.0, https://commons.wikimedia.org/w/index.php?curid=6911957.

Lesson One
Lemon quartz crystal: Photo by Heidi Noggle, 9; Quartz crystal: Photo by Heidi Noggle, 9; Gypsum occurring as blades: Photo by Heidi Noggle, 11; Gypsum rosettes: Photo by Heidi Noggle, 11; Caprock Canyon State Park: By Fredlyfish4 - Own work, CC BY-SA 3.0, https://commons.wikimedia.org/w/index.php?curid=31381977, 11; Alabaster vase: https://commons.wikimedia.org/wiki/File:Alabaster_kohl_pot_HARGM3666.JPG, 12;
Rhombohedral calcite: "Calcite-tch21c" by Rob Lavinsky, iRocks.com – CC-BY-SA-3.0. Licensed under CC BY-SA 3.0 via Commons - https://commons.wikimedia.org/wiki/File:Calcite-tch21c.jpg#/media/File:Calcite-tch21c.jpg, 12; Identification pictures - First picture: https://commons.wikimedia.org/wiki/File:Amethyst._Magaliesburg,_South_Africa.jpg, 13;

Lesson Two
The Periodic Table: Public Domain, By Offnfopt - Own work, Public Domain, https://commons.wikimedia.org/w/index.php?curid=62296883.

Lesson Three
Identification photos: First row, pictures 1,2; Second row, picture 3; Fourth row, pictures 1, 3; Fifth row, pictures 1,2,3; Sixth row, pictures 1,2, by Heidi Noggle, 20;

Lesson Four
Polarizing microscope: "Thin sections" by Woudloper - Own work. Licensed under CC BY-SA 1.0 via Commons - https://commons.wikimedia.org/wiki/File:Thin_sections.JPG#/media/File:Thin_sections.JPG, 24; Granodiorite section: "Granodiorite pmg ss 2006" by No machine-readable author provided. Siim assumed (based on copyright claims). - No machine-readable source provided. Own work assumed (based on copyright claims).. Licensed under CC BY-SA 3.0 via Commons - https://commons.wikimedia.org/wiki/File:Granodiorite_pmg_ss_2006.jpg#/media/File:Granodiorite_pmg_ss_2006.jpg, 24; Gabbro section: "Gabbro pmg ss 2006" by No machine-readable author provided. Siim assumed (based on copyright claims). - No machine-readable source provided. Own work assumed (based on copyright claims).. Licensed under CC BY-SA 3.0 via Commons - https://commons.wikimedia.org/wiki/File:Gabbro_pmg_ss_2006.jpg#/media/File:Gabbro_pmg_ss_2006.jpg, 24; Mineral assemblage of plutonic and volcanic rocks: Mineralogy igneous rocks EN" by Woudloper - Own work. Licensed under Public Domain via Commons - https://commons.wikimedia.org/wiki/File:Mineralogy_igneous_rocks_EN.svg#/media/File:Mineralogy_igneous_rocks_EN.svg, 27; Sandstone: Photo by Heidi Noggle, 31; Coal: Photo by Heidi Noggle, 31; Limestone varieties: photo by Heidi Noggle, 31;

Lesson Five
Willemite: "Willemite 2 w- franklinite zinc orthosilicate Franklin Sussex County New Jersey 1901" by Dave Dyet http://www.shutterstone.com http://www.dyet.com - Own work. Licensed under Public Domain via Commons - https://commons.wikimedia.org/wiki/File:Willemite_2_w-_franklinite_zinc_orthosilicate_Franklin_Sussex_County_New_Jersey_1901.jpg#/media/File:Willemite_2_w-_franklinite_zinc_orthosilicate_Franklin_Sussex_County_New_Jersey_1901.jpg, 35;
Fluorescent minerals: "Fluorescent minerals hg" by (Hgrobe 06:16, 26 April 2006 (UTC)) - credit: Hannes Grobe/AWI - Own work. Licensed under CC BY-SA 2.5 via Commons - https://commons.wikimedia.org/wiki/File:Fluorescent_minerals_hg.jpg#/media/File:Fluorescent_minerals_hg.jpg, 35; Kyanite: By User:Aelwyn - Self made picture, CC BY-SA 3.0, https://commons.wikimedia.org/w/index.php?curid=1002710, 35;Cuprite crystal: "Cuprite-66649" by Rob Lavinsky, iRocks.com – CC-BY-SA-3.0. Licensed under CC BY-SA 3.0 via Commons - https://commons.wikimedia.org/wiki/File:Cuprite-66649.jpg#/media/File:Cuprite-66649.jpg, 35; Turquoise: Public Domain, : "Turquoise.pebble.700pix" by Adrian Pingstone - w:Image:Turquoise.pebble.700pix.jpg. Licensed under Public Domain via Commons - https://commons.wikimedia.org/wiki/File:Turquoise.pebble.700pix.jpg#/media/File:Turquoise.pebble.700pix.jpg, 36; Apatite: "Apatite crystals" by OG59 - Own work. Licensed under Public Domain via Commons - https://commons.wikimedia.org/wiki/File:Apatite_crystals.jpg#/media/File:Apatite_crystals.jpg, 36; Barite: "Baryte Morocco". Licensed under CC BY-SA 3.0 via Commons - https://commons.wikimedia.org/wiki/File:Baryte_Morocco.jpg#/media/File:Baryte_Morocco.jpg, 37; Epsomite: "Epsomite" by WTucker at en.wikipedia. Licensed under CC BY-SA 3.0 via Commons - https://commons.wikimedia.org/wiki/File:Epsomite.jpg#/media/File:Epsomite.jpg, 37;
Sphalerite crystals: https://commons.wikimedia.org/wiki/File:Sphalerite-221270.jpg, 37;
Bromargyrit: "Bromargyrit mit gediegen Silber - San Onofre, Mexiko" by Ra'ike (see also: de:Benutzer:Ra'ike) - Own work. Licensed under CC BY-SA 3.0 via Commons - https://commons.wikimedia.org/wiki/File:Bromargyrit_mit_gediegen_Silber_-_San_Onofre,_Mexiko.jpg#/media/File:Bromargyrit_mit_gediegen_Silber_-_San_Onofre,_Mexiko.jpg, 39; Miersite: "Miersite-215159" di Rob Lavinsky, iRocks.com – CC-BY-SA-3.0. Con licenza CC BY-SA 3.0 tramite Wikimedia Commons - https://commons.wikimedia.org/wiki/File:Miersite-215159.jpg#/media/File:Miersite-215159.jpg, 39; Native gold: "Gold-cat12a" by Rob Lavinsky, iRocks.com – CC-BY-SA-3.0. Licensed under CC BY-SA 3.0 via Commons - https://commons.wikimedia.org/wiki/File:Gold-cat12a.jpg#/media/File:Gold-cat12a.jpg, 41; Native silver: "Silver-280342" by Rob

Lavinsky, iRocks.com – CC-BY-SA-3.0. Licensed under CC BY-SA 3.0 via Commons - https://commons.wikimedia.org/wiki/File:Silver-280342.jpg#/media/File:Silver-280342.jpg, 41;

Lesson Seven

Gypsum (second row, picture 1): "Gypse Arignac" by Didier Descouens - Own work. Licensed under CC BY-SA 4.0 via Commons - https://commons.wikimedia.org/wiki/File:Gypse_Arignac.jpg#/media/File:Gypse_Arignac.jpg, 50; Calcite: First row, picture 3, photo by Heidi Noggle, 50; Fluorite: Picture 1, photo by Heidi Noggle, 51; Apatite: Photos by Heidi Noggle, 51; Feldspar: First row, picture 2, photo by Heidi Noggle, 51 Orthoclase feldspar crystal: "OrthoclaseBresil" by Didier Descouens - Own work. Licensed under CC BY-SA 4.0 via Commons - https://commons.wikimedia.org/wiki/File:OrthoclaseBresil.jpg#/media/File:OrthoclaseBresil.jpg, 52; Albite feldspar crystal: "Albite2" by Didier Descouens - Own work. Licensed under CC BY-SA 3.0 via Commons - https://commons.wikimedia.org/wiki/File:Albite2.jpg#/media/File:Albite2.jpg, 52; Anorthite feldspar crystal: "Anorthite-221029" by Rob Lavinsky, iRocks.com – CC-BY-SA-3.0. Licensed under CC BY-SA 3.0 via Commons - https://commons.wikimedia.org/wiki/File:Anorthite-221029.jpg#/media/File:Anorthite-221029.jpg, 52; Quartz: Pictures 2,3,5, photos by Heidi Noggle, 52; Corundum: Photos by Heidi Noggle, 53; Barite: First row, photos by Heidi Noggle, 54; Gold: Photo by Heidi Noggle, 56; Pyrite: Photo by Heidi Noggle, 56; Limonite: Photo by Heidi Noggle, 56; Quartz: Photo by Heidi Noggle, 57;

Lesson Eight

Egg carton: Public domain, found publicdomainpictures.net, 60;

Lesson Nine

Fluorescent minerals: "Fluorescent minerals hg" by (Hgrobe 06:16, 26 April 2006 (UTC)) - credit: Hannes Grobe/AWI - Own work. Licensed under CC BY-SA 2.5 via Commons - https://commons.wikimedia.org/wiki/File:Fluorescent_minerals_hg.jpg#/media/File:Fluorescent_minerals_hg.jpg, 63; Fluorescent dacite: From picture of author's sample, photographer unknown, 63.

Fluorescent Fish: "Diversity of fluorescent patterns and colors in marine fishes - journal.pone.0083259.g001" by Sparks, J. S.; Schelly, R. C.; Smith, W. L.; Davis, M. P.; Tchernov, D.; Pieribone, V. A.; Gruber, D. F. - Sparks, J. S.; Schelly, R. C.; Smith, W. L.; Davis, M. P.; Tchernov, D.; Pieribone, V. A.; Gruber, D. F. (2014). "The Covert World of Fish Biofluorescence: A Phylogenetically Widespread and Phenotypically Variable Phenomenon". PLoS ONE 9: e83259. DOI:10.1371/journal.pone.0083259.. Licensed under CC BY 4.0 via Commons - https://commons.wikimedia.org/wiki/File:Diversity_of_fluorescent_patterns_and_colors_in_marine_fishes_-_journal.pone.0083259.g001.png#/media/File:Diversity_of_fluorescent_patterns_and_colors_in_marine_fishes_-_journal.pone.0083259.g001.png, 64; Calcite: "Calcite LongWaveUV HAGAM" by This image was provided to Wikimedia Commons by Herbert Art Gallery and Museum, Coventry as part of an ongoing cooperative project. The artefact represented in the image is part of the permanent collection of Herbert Art Gallery and Museum, Coventry.English | македонски | +/−. Licensed under CC BY-SA 3.0 via Commons - https://commons.wikimedia.org/wiki/File:Calcite_LongWaveUV_HAGAM.jpg#/media/File:Calcite_LongWaveUV_HAGAM.jpg "Calcite ShortwaveUV HAGAM" by This image was provided to Wikimedia Commons by Herbert Art Gallery and Museum, Coventry as part of an ongoing cooperative project. The artefact represented in the image is part of the permanent collection of Herbert Art Gallery and Museum, Coventry.English | македонски | +/−. Licensed under CC BY-SA 3.0 via Commons - https://commons.wikimedia.org/wiki/File:Calcite_ShortwaveUV_HAGAM.jpg#/media/File:Calcite_ShortwaveUV_HAGAM.jpg, 64;

Lesson Ten

Information about the diamond industry drawn from: https://www.reuters.com/article/us-angloamerican-debeers-fb/factbox-de-beers-past-and-present-idUSTRE7A33O320111104, and https://en.wikipedia.org/wiki/De_Beers

Lapis lazuli: Photo by "Lapis-lazuli hg" by Hannes Grobe - Own work. Licensed under CC BY-SA 2.5 via Commons - https://commons.wikimedia.org/wiki/File:Lapis-lazuli_hg.jpg#/media/File:Lapis-lazuli_hg.jpg, 69; Tutankamun mask: Photo by "Tutmask" by Jon Bodsworth - http://www.egyptarchive.co.uk/html/cairo_museum_54.html. Licensed under Copyrighted free use via Commons - https://commons.wikimedia.org/wiki/File:Tutmask.jpg#/media/File:Tutmask.jpg, 69;

Pearls: Photo by "Various pearls" by MASAYUKI KATO - MASAYUKI KATO's file. Licensed under CC BY-SA 3.0 via Commons - https://commons.wikimedia.org/wiki/File:Various_pearls.jpg#/media/File:Various_pearls.jpg, 70; Amber jewelry: Photo by "Amber.pendants.800pix.050203". Licensed under Public Domain via Commons - https://commons.wikimedia.org/wiki/File:Amber.pendants.800pix.050203.jpg#/media/File:Amber.pendants.800pix.050203.jpg, 72;

Jet brooch: Photo by "Schwarzer Trauerschmuck2" by Detlef Thomas - de.wikipedia uploaded by Detlef thomas as de:Bild:Trauerschmuck_1.jpg (Eigenes Foto). Licensed under CC BY-SA 2.0 de via Commons - https://commons.wikimedia.org/wiki/File:Schwarzer_Trauerschmuck2.jpg#/media/File:Schwarzer_Trauerschmuck2.jpg, 73;

Iridescent ammonite: Photo by "IridescentAmmonite". Licensed under CC BY-SA 3.0 via Commons - https://commons.wikimedia.org/wiki/File:IridescentAmmonite.jpg#/media/File:IridescentAmmonite.jpg, 74; Ammolite jewelry: Photo by "Ammolite jewellery". Licensed under CC BY-SA 3.0 via Commons - https://commons.wikimedia.org/wiki/File:Ammolite_jewellery.jpg#/media/File:Ammolite_jewellery.jpg, 74; Opalized wood: "Common Opal (Opalized Wood)" by Zuchinni one - Licensed under CC BY-SA 3.0 via Wikipedia - https://en.wikipedia.org/wiki/File:Common_Opal_(Opalized_Wood).png#/media/File:Common_Opal_(Opalized_Wood).png, 75;

Milky opal: "MilkyRawOpal" by CRPeters - Own work. Licensed under CC BY-SA 3.0 via Commons - https://commons.wikimedia.org/wiki/File:MilkyRawOpal.jpg#/media/File:MilkyRawOpal.jpg, 75;

Brazilian opal: "10 7cts Brazilian Crystal Opal" by Daniel Mekis - Own work. Licensed under CC BY-SA 3.0 via Commons - https://commons.wikimedia.org/wiki/File:10_7cts_Brazilian_Crystal_Opal.jpg#/media/File:10_7cts_Brazilian_Crystal_Opal.jpg, 75; Fire opal: "Opledefeu2" by Didier Descouens - Own work. Licensed under CC BY-SA 3.0 via Commons - https://commons.wikimedia.org/wiki/File:Opledefeu2.jpg#/media/File:Opledefeu2.jpg, 75; Opal and diamond pendant: "Rainbow Shield Mintabie Opal Pendant" by Dpulitzer - Own work. Licensed under CC BY-SA 4.0 via Commons - https://commons.wikimedia.org/wiki/File:Rainbow_Shield_Mintabie_Opal_Pendant.jpg#/media/File:Rainbow_Shield_Mintabie_Opal_Pendant.jpg, 75;

Patrick Nurre is from the beautiful state of Montana where as a young boy, he spent many of his Saturdays rockhounding near the Big Horn River. This early interest led him to a lifelong study of the world of geology. His experience has included extensive field study in the Pacific Northwest, the Midwest and Plains states, the Southwestern U.S, and most recently, Israel. Patrick conducts classes and seminars in Seattle, and speaks at numerous homeschool conventions on geology and our young earth. He leads a variety of geology field trips every year, including one to Yellowstone National Park, where he helps families discover the Biblical geology of Yellowstone. Patrick's business, Northwest Treasures, is devoted to producing fine geology specimen kits and curricula from a young earth perspective. Patrick and his wife Vicki have three children and two grandchildren. They live in the Seattle, Washington area.

If you would like to contact Patrick about speaking or field trips:
northwestexpedition@msn.com
For a list of speaking topics: NorthwestRockAndFossil.com

Made in United States
North Haven, CT
05 November 2026

21766163R00046